PERIODIC HAIKU

Jim Bates

Copyright

PERIODIC HAIKU

First Edition. November 25, 2025

Written by Jim Bates

Dedication

For Joe W.

Introduction

The first periodic table to become generally accepted was that of the Russian chemist Dmitri Mendeleev. In 1869, he formulated periodic law as a dependence of chemical properties on atomic mass. As not all elements were then known, there were gaps in his periodic table, and Mendeleev successfully used the periodic law to predict some properties of some of the missing elements. The periodic law was recognized as a fundamental discovery in the late 19th century. It was explained early in the 20th century, with the discovery of atomic numbers and associated pioneering work in quantum mechanics, both ideas serving to illuminate the internal structure of the atom. A recognizably modern form of the table was reached in 1945 with Glenn T. Seaborg's discovery that the actinides were in fact f-block rather than d-block elements. The periodic table and law have become a central and indispensable part of modern chemistry.

The periodic table continues to evolve with the progress of science. In nature, only elements up to atomic number 94 are known to exist; to go further, it was necessary to synthesize new elements in the laboratory. By 2010, the first 118 elements were known, thereby completing the first seven rows of the table; however, chemical characterization is still needed for the heaviest elements to confirm that their properties match their positions. New discoveries will extend the table beyond these seven rows, though it is not yet known how many more elements are possible; moreover, theoretical calculations suggest that this unknown region will not follow the patterns of the known part of the table. Some scientific discussion also continues regarding whether some elements are correctly positioned in the table. Many alternative representations[1] of the periodic law exist, and there is some discussion as to whether there is an optimal form of the periodic table.

p

This collection focuses on the first 118 elements. An element's number on the table is its atomic number, which is the number of protons the element has. The haiku are written in the 5-7-5 syllable format. They are meant to be fun to read. If they spur the reader forward to investigate an element further, so much the better!

1. https://en.wikipedia.org/wiki/Alternative_periodic_tables

Table of Contents

Hydrogen
Helium
Lithium
Beryllium
Boron
Carbon
Nitrogen
Oxygen
Florine
Neon
Potassium
Sodium
Magnesium
Aluminum
Silicon
Phosphorus
Sulfur
Chlorine
Argon
Potassium
Calcium
Scandium
Titanium
Vanadium
Chromium
Manganese
Iron
Colbalt
Nickel
Copper
Zinc
Gallium
Germanium

Arsenic
Selenium
Bromine
Krypton
Rubidium
Yttrium
Zirconium
Niobium
Molybdenum
Technetium
Ruthenium
Rhodium
Palladium
Silver
Cadmium
Indium
Tin
Antimony
Tellurium
Iodine
Xenon
Cassiem
Barium
Lanthanides
Hafnium
Tantalum
Tungsten
Rhenium
Osmium
Iridium
Platinum
Gold
Mercury
Thallium
Lead
Bismuth

Polonium
Astatine
Radon
Francium
Radium
Actinium
Thorium
Protactinium
Uranium
Neptunium
Plutonium
Americium
Curium
Berkelium
Californium
Einsteinium
Fermium
Mendelevium
Nobelium
Lawrencium
Rutherfordium
Dubnium
Seaborgium
Bohrium
Hassium
Meitnerium
Darmstadtium
Roentgenium
Copernicium
Nihonium
Flerovium
Muscovium
Livermorium
Tennessine
Oganesson

Theoretical Elements:

- Ununennium
- Unbinilium
- Unbiunium
- Unbibium
- Unbitium

Hydrogen (1)

At the beginning
Solitary Element
Looking to connect.

Helium (2)

Happy Helium
Two atoms bond side by side
Laughing all the way.

Lithium (3)

Helpful element
Especially when there's stress
Time to mellow out.

Beryllium (4)

Structurally stable
Viewing windows for X-rays
Bonds well with others.

Boron (5)

Think boric acid
Other cleaning compounds for sure
Very versatile.

Carbon (6)

Unique element
Under the weight of eons
Precious diamonds form.

Nitrogen (7)

Abundant in air
Made into fertilizer
Loved by all living plants.

Oxygen (8)

Premier element
Comes from photosynthesis
Making life possible.

Fluorine (9)

Highly reactive
Dangerous fluorocarbons
Handle with much care.

Neon (10)

Second noble gas
Excitable element
Very colorful.

Sodium (11)

Osmotic pressure
Essential in animal cells
Keeps the flow going.

Magnesium (12)

Stellar element
Formed when super novas explode
Unique cosmic dust.

Aluminum (13)

Found in the Earth's crust
Third most common element
Rules the kitchen.

Silicon (14)

Sandy silicon
Mix it up to form concrete
Hard and durable.

Phosphorus (15)

Morning star to Greeks
Essential to all life forms
Makes matches light fast.

Sulphur (16)

Yellow in color
Used in sulfuric acid
Burning sensation.

Chlorine (17)

Room temperature gas
A common disinfectant
Swimming pool odor.

Argon (18)

Lazy to the Greeks
Abundant yet inactive
Used in some lasers.

Potassium (19)

Silver white metal
Easily cut by a knife
Living cells need it.

Calcium (20)

Alkaline metal
Found in limestone deposits
Sealife fossils too.

Scandium (21)

Rare Earth element
Bonds well with aluminum
Used to build airplanes.

Titanium (22)

Transition metal
Beautiful lustrous color
Incredibly hard.

Vanadium (23)

Strong when oxidized
Used as a steel additive
Compounds are toxic.

Chromium (24)

Twenty-four protons
For its atomic number
Makes many pigments.

Manganese (25)

Brittle silvery metal
Important in our diet
Good for bone structure.

Iron (26)

Most common by mass
Known about since ancient times
Health necessity.

Cobalt (27)

Cobalt blue pigment
Used in paints and by jewelers
Artist's element.

Nickel (28)

Silver white metal
Greatest use is stainless steel
Worth more than five cents.

Copper (29)
Native mineral
Used directly from the ground
Nice green patina.

Zinc (30)

An alloy in brass
Essential trace element
Nutrient for coral.

Gallium (31)

Common in alloys
Even in some thermometers
Very conductive.

Germanium (32)

Used in transistors
Similar to silicon
Binds with oxygen.

Arsenic (33)

Strong alloy with lead
Mystery writers love it
Plays well with old lace.

Selenium (34)

Found in sulfide ores
Used in semiconductors
Also glassmaking.

Bromine (35)

Reddish brown substance
Liquid at room temperature
Very volatile.

Krypton (36)

Inert noble gas
Lighting and photography
Fluorescent lamps.

Rubidium (37)

Soft white/grey solid
Give purple to fireworks
Found in atomic clocks.

Strontium (38)

Many isotopes
Gives fireworks red color
Early TV tubes.

Yttrium (39)

Rare earth element
Named for village in Sweden
Old cathode ray tubes.

Zirconium (40)

Greyish white color
High temperature usage
Transition metal.

Niobium (41)

Greek Mythology
Alternative to nickel
Can be distorted.

Molybdenum (42)

Used in steel alloys
Comes from Greek word meaning lead
Found in minerals.

Technetium (43)

Fission by-product
Radioactive metal
Occurs in Earth's crust.

Rubidium (44)

Alkali metal
Is easily vaporized
Not very abundant.

Rhodium (45)

Found in nickel ores
Catalytic convertors
Hard, durable metal.

Palladium (46)

Ore deposits rare
Catalytic converters
Silver white metal.

Silver (47)

Highly conductive
From Latin word Argentum
Low toxicity.

Cadmium (48)

Resists corrosion
Soft malleable metal
Not flammable.

Indium (49)

Many isotopes
A very soft element
Flat panel displays.

Tin (50)

Used to plate steel
Highly conductive metal
Some yellow pigments.

Antimony (51)

Known since ancient times
Alloys with lead and tin
In flame retardants.

Tellurium (52)
Mildly toxic
Is used in solar panels
Rare in the earth's crust.

Iodine (53)
Found in many forms
An essential mineral
Violet at high temp

Xenon (54)

Dense colorless gas
Propellant in some spacecraft
Used flash strobe lamps.

Cesium (55)

In atomic clocks
Liquid metal near room temp
Used for drilling.

Barium (56)

Very reactive
Green color in fireworks
X-ray imaging.

Lanthanum (57)

Has many uses
First in Lanthanoid series
Additive in glass.

Cerium (58)

Soft whitish metal
Used to decolorize glass
Vivid red color.

Praseodymium (59)

Silvery color
Yellow color to glasses
Pliable metal.

Neodymium (60)

Popular additive
Is slightly malleable
Is used in glass dyes.

Promethium

Rare, rare element
From spontaneous fusion
Radioactive.

Samarium (62)

Rare Earth element
Main use is cobalt magnets
Very volatile.

Europium (63)

Can cut with a knife
Rarest rare Earth element
Red in TV screens.

Gadolinium (64)

Used in some alloys
In medical imaging
Silver white metal.

Terbium (65)

Is non-hazardous
Found in many minerals
In sonar systems

Dysprosium (66)

Bright silver luster

Used in electric motors

Mildly toxic..

Holmium (67)

Relatively rare
Extracted from Monazite
Fairly corrosive.

Erbium (68)

Is mined in Sweden
Used in laser surgery
Silvery metal.

Thulium (69)

Silver grey luster
Easy metal to work with
Rare and costs a lot.

Ytterbium (70)

Quite malleable
Mined in Ytterby Sweden
In atomic clocks.

Lutetium (71)

Silver white color
In nuclear medicine
Resists corrosion.

Hafnium (72)

Lustrous silver grey
Produced from zirconium
Used in power plants.

Tantalum (73)

Named for Tantalus
Found in mineral groups
Quantum processors.

Tungsten (74)

A high melting point
Naturally in Earth's crust
In many alloys.

Rhenium (75)

High melting point
Is very rare in Earth's crust
Named for river Rhine.

Osmium (76)

Brittle blueish white
Trace found in platinum ores
Fountain pen nib tips.

Iridium (77)

Hard very brittle
In high-performance spark plugs
In meteorites.

Platinum (78)

Dense malleable
Found with nickel and iron
Is non-corrosive.

Gold (79)

Soft malleable
On Earth from meteorites
Often grey color.

Mercury (80)

Heavy and silver
Gives red pigment vermillion
Toxic concerns.

Thallium (81)

Silvery and white
Is not found free in nature
Used in rat poison.

Lead (82)

Dense heavy metal
Used since prehistoric times
Can bond with itself.

Bismuth (83)

Known since ancient times
Naturally occurring
Alloy replaces lead.

Polonium (84)

Very rare metal
Highly radioactive
Very dangerous.

Astatine (85)

Very short half-life
Rare natural element
Radioactive.

Radon (86)

Radioactive
Formed by radium decay
Lots of it around.

Francium (87)

Very unstable
Very radioactive
Has never been seen.

Radium (88)

Is very toxic
Very radioactive
Luminesces paint.

Actinium (89)

Soft silvery white
First in actinoid series
Radioactive.

Thorium (90)

Light silver metal
Weakly radioactive
Named for Norse God Thor.

Protactinium (91)

Very reactive
Found in spent nuclear fuel
Radioactive.

Uranium (92)

Named for Uranus
In nuclear power plants
Handle with thick gloves.

Neptunium (93)

Named after Neptune
Spacecraft electricity
Radioactive.

Plutonium (94)

Named after Pluto
In nuclear reactors
Occurs in nature.

Americium (95)

Radioactive
Produced in 1944
In smoke detectors.

Curium (96)

Think Marie Curie
Very radioactive
Very long half-life.

Berkelium (97)

Named after Berkeley
Soft white radioactive
Very little made.

Californium (98)

Silver white metal
Easily cut with a knife
Used in reactors.

Einsteinium (99)

1952

Formed from first hydrogen bomb

Radioactive’

Fermium (100)

Very fast decay
From nuclear fallout
1952.

Mendelevium (101)

Small quantities
From nuclear bombardment
Named for Mendeleev.

Nobelium (102)

Twelve isotopes
Named after Alfred Nobel
Very short half-life.

Lawrencium (103)

For Ernest Lawrence
Invented the cyclotron
Fourteen isotopes.

Rutherfordium (104)

Radioactive
Named nineteen ninety-seven.
Ernest Rutherford.

Dubnium (105)

Very short half-life
Named for the town of Dubna
Limited research.

Seaborgium (106)

Not found in nature
Named for Glenn Seaborg
1974.

Bohrium (107)

Named after Niels Bohr
A synthetic element
Not found in nature.

Hassium (108)

Heavy element
Very radioactive
Comes from cold fusion.

Meitnerium (109)

Named for Lise Meitner
Very radioactive
1982.

Darmstadtium (110)

Very unstable

Very radioactive

1994.

Roentgenium (111)

Radioactive
Is made in the lab
Wilhelm Roentgen.

Copernicium (112)

Radioactive
Half-life of thirty seconds
1996.

Nohnium (113)

Ten-second half-life
A synthetic element
Made in small amounts.

Flerovium (114)

Dubna Russia lab
Very radioactive
1999.

Moscovium (115)

From Dubna Russia
A synthetic chemical
In 2003.

Livermorium (116)

Not found in nature
A synthetic element
Radioactive.

Tennessine (117)

Synthetically made
Named for state of Tennessee
Very volatile.

Organesson (118)

2016

Yuri Organessian

Russian physicist.

Author's Note:

An extended periodic table theorizes about chemical elements[1] beyond those currently known and proven. The element with the highest atomic number[2] known is oganesson[3] (118), which completes the seventh period[4] (row) in the periodic table[5]. All elements in the eighth period and beyond thus remain purely hypothetical.

They are:

Ununennium (119)
Unbinilium (120)
Unbiunium (121)
Unbibium (122)
Unbitium (123)

1. https://en.wikipedia.org/wiki/Chemical_elements
2. https://en.wikipedia.org/wiki/Atomic_number
3. https://en.wikipedia.org/wiki/Oganesson
4. https://en.wikipedia.org/wiki/Period_(periodic_table)
5. https://en.wikipedia.org/wiki/Periodic_table

Acknowledgments

I dedicated this collection to author JB Wocoski. His collection of haiku, *Wakizashi Haiku*, is a fantastic read. Check it out if you get a chance.

Many thanks go out to my friend and mentor, PC Darkcliff, author of the fabulously exciting fantasy/adventure series "Shattered Empire" (among many other stunning books). PC, your support over the years has been and continues to be incredible.

Also, thanks go out to the members of the Facebook group TRENDING POETS for all of your support. I appreciate it more than I can say.

Finally, thank you, Andjela Vujic, for another great cover! It's perfect.

BOOKS PUBLISHED BY JIM

Here are books published by Jim. They are available at most traditional outlets.

Dear Reader. Please leave a review of the book you just finished or any of my other books. It will be much appreciated!

Biography

Jim lives in a small town in Minnesota. He loves to write! His stories and poems have appeared in over five hundred online and print publications. His collection of short stories, *Resilience,* is published by Bridge House Publishing. *Short Stuff,* a collection of flash fiction and drabbles, is published by Chapeltown Books. *Periodic Stories, Periodic Stories Volume Two, Periodic Stories Volume Three – A Novel,* and *Periodic Stories Volume Four* are published by Impspired. *Dreamers,* a collection of short stories, is published by Clarendon House Publishing. *Something Better,* a dystopian adventure novella, and the novel *The Alien of Orchard Lake* are published by Dark Myth Publications. In the fall of 2022, his collection entitled *Holiday Stories* was published by Impspired as was his collection of poetry, *Haiku Seasons*. In February 2023, a collection of flash fiction, *Dancing With Butterflies,* was published by Impspired. In July 2023, his YA novella *The Battle of Marvel Wood* was published by Impspired. In July 2023, his collection of poems, *The Alchemy of Then,* was published by Impspired. In August 2023, his novel *Conversations With the Dead* was published by Impspired. In September 2023, his collection of poems, *The Metaphysics of Now,* was published by Impspired. In October 2023, his collection of short *stories, Old Man Jasperson and Other Stories,* was published by Bridge House Publishing. In January 2024, his collection of poems, *The Evolution of Tomorrow,* was published by Impspired. In January 2024, his novella *Eye of the Beholder* was published by Dark Myth Publications. In March 2024, his collection of poems, *The Science of Forever,* was published by Impspired. In August 2024, his collection of long short stories, *The Stargazer and Other Stories,* was published by Jim. In November 2024, his collection of long short stories, *On Rainy Lake and Other Stories,* was published by Jim. In January 2025, his YA adventure/fantasy novella

The Time Traveling Healer was published by Jim. In March 2025, his YA adventure/dystopian novel *The Defectives* was published by Jim. In May 2025, his YA adventure/dystopian novel *The Gallantiers* was published by Jim. In August 2025, his YA adventure/dystopian novella *The Two Tribes* was published by Jim. In October 2025, his YA adventure/dystopian novel *The Explorers* was published by Jim. It's the 4th book in his "Creekside Chronicle" series. In September 2025, his collection of poems, *The Haiku Year,* was published by Jim. His short story "Aliens" was nominated by The Zodiac Press for the 2020 Pushcart Prize. His story "The Maple Leaf" was voted 2021 Story of the Year for Spillwords. He was voted December 2022 Author of the Month for Spillwords. He also reads his stories for Tony Cranson and *Talking Stories Radio,* and for *Jim's Storytime* and *Jim's Poetry Corner* on his website. He lives in a small town west of Minneapolis, Minnesota. All of his work can be found on his blog at www.theviewfromlonglake.com[1]

1. http://www.theviewfromlonglake.com

www.ingramcontent.com/pod-product-compliance
Lightning Source LLC
LaVergne TN
LVHW101616210926
848480LV00009B/575